UNDERWATER FACTS

©2024
BookLife Publishing Ltd.
King's Lynn, Norfolk
PE30 4LS, UK

All rights reserved.
Printed in India.

A catalogue record for this book is available from the British Library.

ISBN: 978-1-80505-691-1

Written by:
Rod Barkman

Edited by:
Rebecca Phillips-Bartlett

Designed by:
Amy Li

FSC
www.fsc.org
MIX
Paper | Supporting responsible forestry
FSC® C195953

All facts, statistics, web addresses and URLs in this book were verified as valid and accurate at time of writing. No responsibility for any changes to external websites or references can be accepted by either the author or publisher.

AN INTRODUCTION TO BOOKLIFE RAPID READERS...

Packed full of gripping topics and twisted tales, BookLife Rapid Readers are perfect for older children looking to propel their reading up to top speed. With three levels based on our planet's fastest animals, children will be able to find the perfect point from which to accelerate their reading journey. From the spooky to the silly, these roaring reads will turn every child at every reading level into a prolific page-turner!

CHEETAH
The fastest animals on land, cheetahs will be taking their first strides as they race to top speed.

MARLIN
The fastest animals under water, marlins will be blasting through their journey.

FALCON
The fastest animals in the air, falcons will be flying at top speed as they tear through the skies.

PHOTO CREDITS Images are courtesy of Shutterstock.com. With thanks to Getty Images, Thinkstock Photo and iStockphoto. Recurring – Milano M, MaryDesy, Nastya Vaulina, Baskiabat. Cover – BlueRingMedia, Macrovector, Gear Digital, Bohdan Populov, KatyGr5, MrVettore, KittyVector. 4–5 – Baksiabat, NotionPic, Pogorelova Olga, Willyam Bradberry. 6–7 – David Herraez Calzada, GN.Studio, Lucia.Pinto, Maquiladora. 8–9 – Dacian Galea, Morphart Creation. 10–11 – Fer Gregory, mentalmind. 12–13 – Catmando, Iconic Bestiary, Rebecca Schreiner. 14–15 – BlueRingMedia, ijimino, Invision Frame, moj0j0. 16–17 – divedog, Flash Vector, Mascha Tace, NotionPic. 18–19 – alazur, Arthur Balitskii, BlueRingMedia, Oliver Denke, Lubenica. 20–21 – Craig Lambert Photography, MicroOne, nastyaartnik, Wise ant. 22–23 – Alexandre.ROSA, Kolonko, Net Vector. 24–25 – Stas Moroz, wickerwood. 26–27 – Artemii Sanin, Evgeniy yes, Lillac. 28–29 – Julia Faranchuk, Jyggalag, Sabelskaya. 30 – Sabelskaya.

CONTENTS

PAGE 4	Totally Extreme Underwater
PAGE 6	Extreme Diving
PAGE 8	Extreme Underwater Inventions
PAGE 10	Underwater Treasure
PAGE 12	Extreme Underwater Animals
PAGE 16	Extreme Pressure
PAGE 18	The Mariana Trench
PAGE 20	Underwater Sound
PAGE 22	Underwater Volcanoes
PAGE 24	Underwater Exploration
PAGE 26	The Ocean Zones
PAGE 30	EXTREME Underwater
PAGE 31	Glossary
PAGE 32	Index

WORDS THAT LOOK LIKE THIS ARE EXPLAINED IN THE GLOSSARY ON PAGE 31.

TOTALLY EXTREME UNDERWATER

Our world is covered in water. In fact, about 71 <u>percent</u> of the Earth is covered in water.

Under that water is a whole world of amazing and EXTREME things. From huge fish to EXTREME mysteries, our oceans are packed with the unknown.

What we do know about this underwater world is surprising... shocking... EXTREME!

So, take a deep breath. Hold it. Now, break through the surface and dive down deep into an ocean of EXTREME facts!

EXTREME DIVING

Speaking of diving...

The world's deepest dive is 332.35 metres. It was carried out by Egyptian diver, Ahmed Ahdel Gahr in 2014.

That is nothing compared to the emperor penguin, which can dive down to over **475** metres without any equipment.

SCUBA means Self-<u>Contained</u> Underwater Breathing <u>Apparatus</u>. This is the name of the equipment used to dive.

SCUBA GEAR

The longest scuba dive lasted **145** hours and **25** minutes!

EXTREME UNDERWATER INVENTIONS

All this diving had to start somewhere. But where and when?

One of the first working underwater suits was invented by a French man in **1715**. Air was fed through a tube to the suit.

The Necker Nymph is the first underwater plane. Pilots wear scuba gear and the plane flies underwater!

The diving bell was an EXTREME way to go underwater in the 1600s. It trapped air in a wooden bell shape that someone would stand in.

UNDERWATER TREASURE

An EXTREME amount of treasure has been lost at sea.

It is thought that there is over £47 billion in treasure at the bottom of the ocean.

EXPERTS SAY THAT THEY CANNOT KNOW FOR SURE, BUT THIS IS THEIR BEST GUESS.

The most <u>valuable</u> underwater treasure found was worth around £314 million.

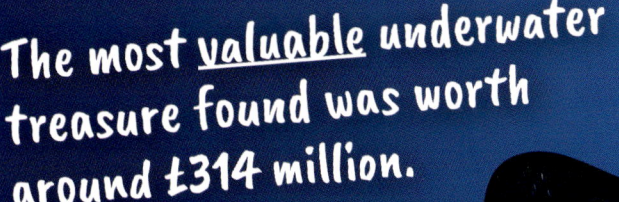

Treasure hunters are still looking for the wreck of the Flor de la Mar. It is thought to have a treasure worth over £2 billion hidden in it.

EXTREME UNDERWATER ANIMALS

There are more animals than we know about underwater. Some are quite EXTREME!

The immortal jellyfish does not like to die. If injured or starving, it can turn itself back into a baby and start its life cycle all over again!

The coelacanth fish was thought to have died off millions of years ago. In **1938**, a living one was discovered.

The largest animal on Earth is found under the water: the blue whale. They can grow to over 30 metres long!

An electric eel can put out enough power to light ten lightbulbs!

There is a creature that swims into other fish's mouths and attaches themselves to their tongues. This makes the real tongue fall off! Extremely yuck!

There have been <u>legends</u> of giant squids for thousands of years. In 2012, a live giant squid was filmed for the first time.

Cuvier's beaked whales dive deeper than any other air-breathing animal. They can dive to two thousand, nine hundred and ninety-two metres.

EXTREME PRESSURE

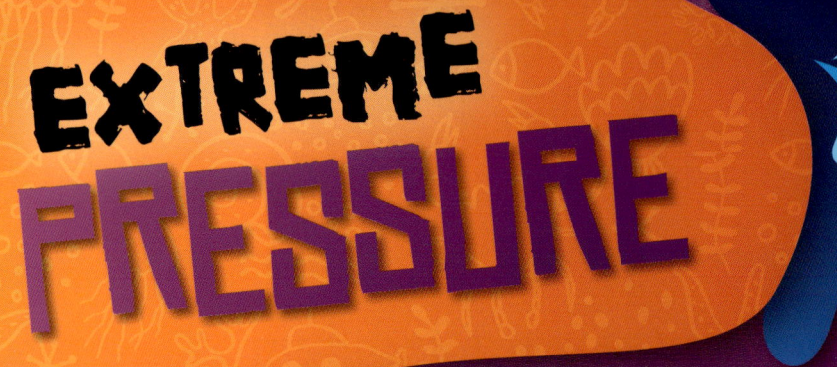

The deeper underwater you go, the more the weight of the water presses on you. This is called pressure. How EXTREME can this pressure get?

The pressure at the deepest part of the ocean is equal to the weight of 48 jumbo jets balanced on your head!

Even if you could withstand the weight of the water, that EXTREME pressure changes how the water in your body acts.

Before EXTREME pressure crushes you, it will do some strange things to the inside of your body!

THE MARIANA TRENCH

The Mariana Trench is the deepest place on Earth.

It is more than two thousand, five hundred and fifty kilometres long.

It has an average width of 69 kilometres.

It is eleven thousand and thirty-four metres deep.

If you put Mount Everest, the highest point on Earth, in the Mariana Trench, its peak would still be two thousand, one hundred and thirty-three metres under the water's surface!

The deepest part of the trench is called the Challenger Deep.

UNDERWATER SOUND

Believe it or not, sound travels better through water than it does through air!

Sound travels at an EXTREME speed underwater — four times faster than it does through air!

Sound actually travels faster in the deepest parts of the ocean!

The blue whale is one of the ocean's loudest animals.

A blue whale's song can reach **180** decibels. That is as loud as a jet plane!

In 1997, a mysterious sound was recorded underwater. It was called the Bloop.

UNDERWATER VOLCANOES

Most of Earth's volcanoes are hidden deep under the ocean.

There are over one million underwater volcanoes.

Water cools lava faster than air. This fast cooling means that underwater volcanoes often make volcanic glass. This is known as obsidian.

The lava from underwater volcanoes can form islands.

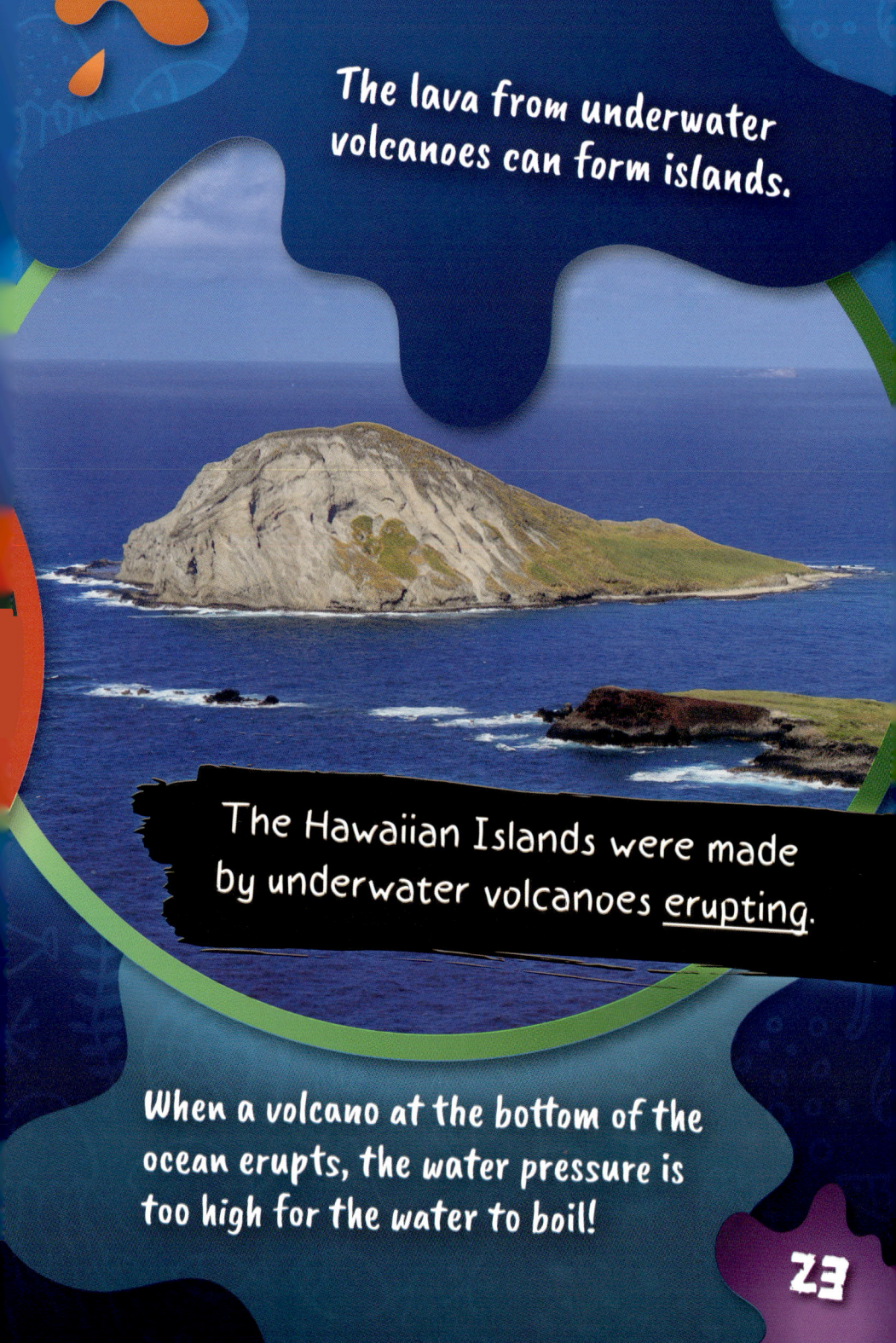

The Hawaiian Islands were made by underwater volcanoes <u>erupting</u>.

When a volcano at the bottom of the ocean erupts, the water pressure is too high for the water to boil!

UNDERWATER EXPLORATION

Most of the Earth covered in water, so there is a lot to explore under the waves.

Less than five percent of the Earth's oceans have been explored!

More people have walked on the Moon than on the floor of the Mariana Trench.

The ocean floor is being mapped using sound.

Scientists make sound maps by bouncing sound off the ocean floor and measuring how soon it comes back.

This is called SONAR. It stands for Sound Navigation and Ranging.

THE OCEAN ZONES

The ocean has five zones. Each one is different and EXTREME in its own way.

The sunlight zone is the area from the surface to 200 metres deep.

This is the zone that sunlight reaches. Most life is found here.

26

The twilight zone starts at 200 metres deep and reaches 1,000 metres deep.

There is almost no <u>visible</u> light this deep, so there are almost no plants in the twilight zone.

The midnight zone reaches to around 4,000 metres deep.

The midnight zone is freezing cold and pitch black!

The abyssal zone reaches to 6,500 metres deep.

Many of the animals that live this deep glow in the dark! This is called bioluminescence.

The hadal zone is the deepest depths of the ocean.

Only ocean trenches, such as the Mariana Trench, make up the hadal zone.

Under EXTREME cold, EXTREME darkness and EXTREME pressure, the hadal zone is home to over **400** known <u>species</u>.

EXTREME UNDERWATER

Under the surface of the water lies a world almost alien to us. It is filled with amazing creatures, lost treasures and EXTREME dangers.

We may never explore all of it, but there are many more EXTREME underwater facts to learn!

GLOSSARY

APPARATUS equipment or machinery needed for a particular purpose

CONTAINED held within

DECIBELS the units used to measure how loud a sound is

EQUIPMENT things made or used for a particular activity

ERUPTING breaking out with force

EXTREME at the highest level

LEGENDS stories from a long time ago that have been passed down through generations

PERCENT one part in every 100

SPECIES a group of very similar animals or plants that can produce young together

VALUABLE worth a lot of money or considered important and useful

VISIBLE able to be seen by the human eye

INDEX

DIVING 6–9, 15

EQUIPMENT 6–7

FISH 4, 13–14

HAWAIIAN ISLANDS 23

LAVA 22–23

SONAR 25

SOUND 20–21, 25

TREASURE 10–11, 30

VOLCANOES 22–23

WHALES 13, 15, 21